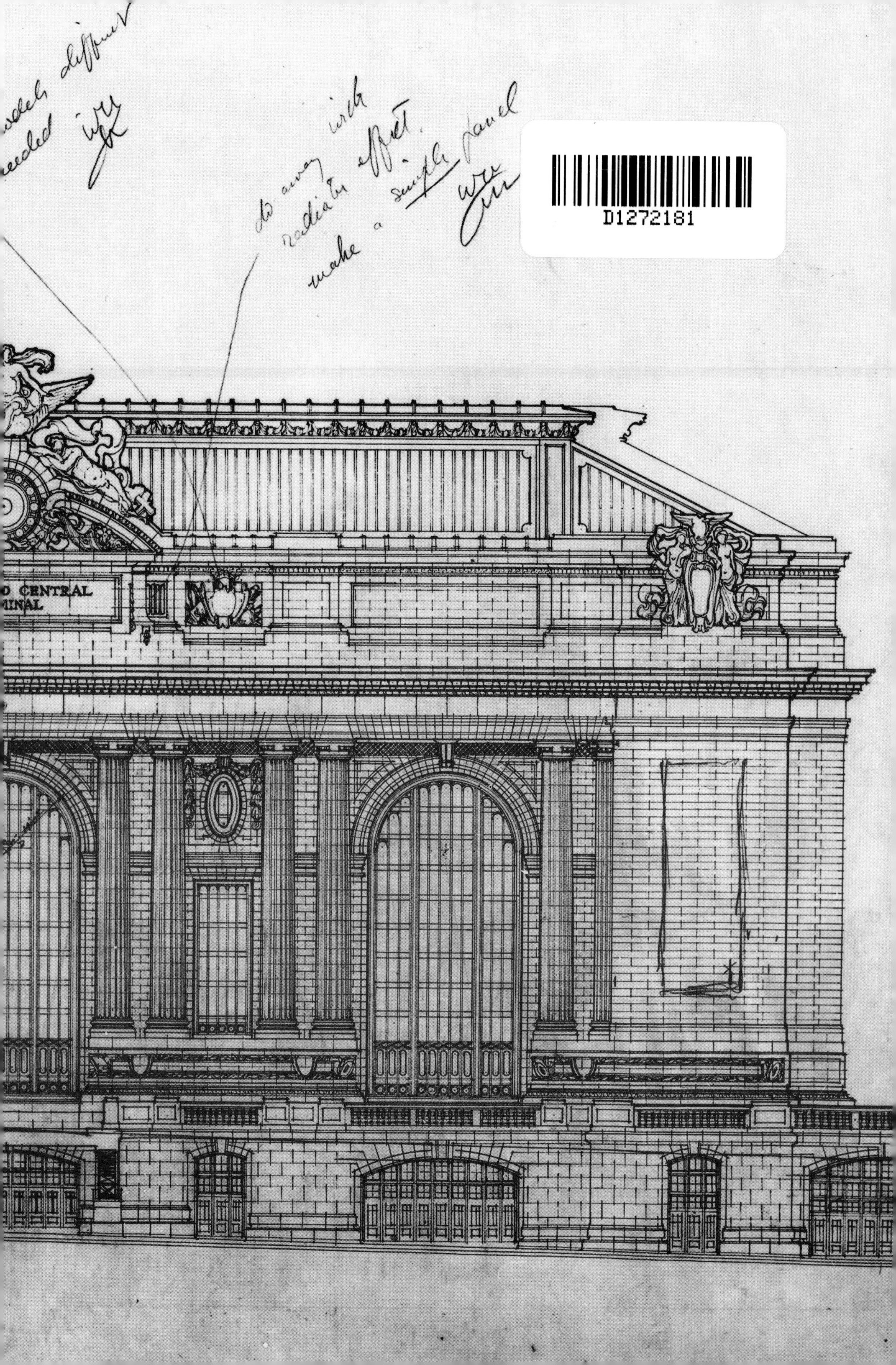
do away with radiator effect. make a simple panel
O CENTRAL
MINAL

GRAND CENTRAL TERMINAL

GRAND CENTRAL